MW00345137

belongs to Rachel

*contents | convents*

urf

It saves wear and tear of muscle and disposition, lessens the production of domestic quarrels, adds to the pleasure and satisfaction of living…If it not be a blessing to humanity, the fault lies with the folks and not the stuff.

> —John McLaurin, *Sketches in Crude Oil: Some Accidents and Incidents of the Petroleum Development in all Parts of the Globe* (1896)

Man invented the machine in order to discover himself
Yet I have heard a lady say "Il fait l'amour comme
une machine à coudre," with no inflection of approval.

> —Mina Loy, "The Oil in the Machine?" (1921)

There on the road was a mangled corpse—a ground squirrel had tried to cross, and a car had mashed it flat; other cars would roll over it, till it was ground to powder and blown away by the wind.

> —Upton Sinclair, *Oil!* (1927)

Created from the lush vegetation and animal fats of the Carboniferous and adjoining periods, holding in itself the black essence of all life that had ever been, constituting in fact a great deep-digged black graveyard of the ultimate eldritch past with blackest ghosts, oil had waited for hundreds of millions of years, dreaming its black dreams, sluggishly pulsing beneath Earth's stony skin, quivering in lightless pools roofed with marsh gas and in top-filled rocky tanks and coursing through myriad channels and through spongy rocky bone, until a being evolved on the surface with whom it could realise and expend itself.

> —Fritz Leiber, *The Black Gondolier* (1964)

I'll run out on the way; the gauge has been warning me for quite a while that the tank is in reserve. They have been warning us for quite a while that underground global reserves can't last more than twenty years or so.

> —Italo Calvino, "The Petrol Pump" (1974)

Corlie: "Why do men have to fight?"
Hunter: It's always been about who gets more."
Corlie: "It has to change."

> —*Warlords of the Twenty-First Century/Battletruck* (1982)

*the concrete tower had been a foundry,*
*an animal feed supplement factory, and a propone distributor*

*carbon sink*

The following cut-up-essays-islands began as a long poem about the petropastoral, a sump excited by a strip of yellow plastic, snarled in the branches of a maidenhair tree outside my window

each morning this tree waved a sheer sea at me, once seared by hot air and pressure, a memory of an old rock, it bruised an asking to excavate the elastic networks between my organs and plastic to stay the shared molecular architecture between ghostly carbonauts

to scatter flares of airy industry

yeah it's a tough oil world out there

I spent a year misreading the wildcats: concussive leavings, misseeings, malapropisms, rereadings, splurtings of a variety of textual sources from newspaper articles in Nu Zild to nineteenth century books, to drill vocabularies that verbify slow violence into new speeds, to create textual gyres that mimic the new soup islands, to stretch the entanglements of fossil fuels, PA fracking, oceanic chemistry, climate change, human and nonhuman carbon-based cosmonauts, the lonely petronauts of planet urf

I, a smut hound, unurfed questions: how to write a sustainable encounter with dead sea monsters and plastic without fabricating disgust?

how to navigate unseen ecologies with microscopic life forms seen suspended only in underwater photographs?

how to document the multiversal mixing of plastics and the petrosublime, the natural history of plastispheres in the Great Pacific Garbage Patch?

how to carbonise poetry?

how to blowout, pipe, leak, dredge, spill the desires of language, mine the encyclopedic spoor, the toxic fossils buried in texts (please note how the geo-rhetorics seismic so easily on the lips)?

okay. I have a frog called plastic in my throat

I fantasise about the inverse possibilities of oily writing

can poetry become a carbon sink, absorbing excess $CO_2$ instead of releasing it through performance or publication?

how to recall the trace of petronauts immigrating into the abyss zones and outerspaces of the Mariana and Kermandac trenches where PCBs soak pale pink amphipods and sea cucumbers, the human dispersing through chemicals in nonhuman waters hostile to our organs, alien ways of breathing

to speak in flinty tongues to organic sea glass that face upwards, to feel light from other alien bodies in uneven darkness?

how to write that poem?

*weeds traced the tracks of railcars*
*that carried coal to Willow Steam Factory*

*garden | eden*

The old petronauts were not native
to the floating islands
but they recalled
the petroleum romances:
the tallow-dip and lard-oil,
pine-knots and smokey candles,
swinging sconces
that lit their prayerbooks better.
In holy texts they rubbed
their tongues with shale
and distended their ripened stomachs.
Ore poured from orefaces,
illuminated caverns with plankton.
One petronaut asked
if the Lord wanted
a thousand rams
or ten thousand rivers of oil,
but he misread the olives.
Even in the Garden of Eden,
Adam the graper
coated the tree with coal oil,
to sour the insects,
but incited the pipeline
snake to slide down
the sticky tree
with astonishment.
Eve—Dirty Gertie
could have avoided
trouble if you had bit into
the flinty rock instead

*the disused smokestack of the powerhouse was a cenotaph*

*blowout | blowter*

The ancient petronauts also dug asphaltum
at the sites of Sodom and Gomorrah,
on the plain north of the dead sea,
where the Tel Aviv Stock Exchange
claimed an oil reserve worth $320 million.
Petrotourists spread bad reviews that
the slime-pit gardens blackened
the flinty darlings of Sodom
but it's likely the locals were
simply inhospitable to strangers
who bought licenses to the pits.
As a maxim of failing,
the blowout between the petronauts
spilled 800 barrels of
viscous rock into the Jordan River
wiping out aquatic life and birds,
and prompting government officials
to close the waterway for public swimming,
boating, wading, and fishing
although an expert later
struggled to distinguish the
thick dark mud from bloom.
Official reports noted only
a tiny sheen on the water
even as the blossoms of bitumen
under the cities' foundations
fuelled the eternal fire

*the petronaut listened:*
*a melody of trains had fossilised the wind*

*garble | gargle*

The poetronauts also misread
the slow blowouts
spun devotionals
to sticky wildernesses
recycling love affairs
with petroplastics
into time-stamped disgust.
Oil isn't spiritual.
Plastic isn't the poem of our time.
Plastic bags are useful
for people in transient,
food for whales and gulls.
Plastic bags are often found in whale stomachs
who misread the sheer luminance for jellyfish.
Liquid flows through a sponge like
the middle of a poem.
*Oil mixes with seawater*
*and forms an emulsion like mousse,*
*left out, the surface crusts over*
*but the inside still has*
*the consistency of mayonnaise.*
How many gallons of crude oil are
in this poem?
How many gallons of crude oil
from the 2011 floundering of Rena
in this poem?
How many gallons of crude oil
to help the urf with birds?
How many gallons of crude oil
to mask the waxy
faces of inmates rented
to clean up sea lions,
to be plankton
somewhere between solid and wet?
How many gallons of crude oil

to feel ease
in sheens togethered
by water blooms?

Rivers of oil.
What kind of poem is this?

Mayonnaise

*carbon sink*

March 12, 2017

A tree's list of ocean plastic:

fish plaiting rayon thread
sud currents fill viney frill
crabs pinching pen caps
nurdle cud
and polyester shrimp
oar feet
gelatinous jelly bags
venus flower baskets
pink vase
or toothpaste tubing
sea gooseberry
and brittle star
braining beads with
bottled spookfish
cholera commas
Japanese petronauts, who discovered bacteria burrow into pits of ocean plastic, gorging themselves on melted
cells of polythene surfaces like *hot barbecue briquette thrown into snow*
meal moths
and wax moths wearing
tattered pants made with ultrasuede
and tar balls
glad wrap anemone
poems like islands without bridges or causeways
(that is, you have to *work* to get *there*. Trust me, there's a metronome to my magpie)

also inhaled questions that are not questions:
how to love a virus with concussive tenderness
how to scale eroticism to two chromosomes, which colonise the soupy islands
(we're caught between a romance and a hard plaint)
the islandary vs. the planetary
capitalism: the fine art of terraforming seas
and feeling empathy for bottle caps bursting with dead sea birds

*oil colonised the mouth until the river was mute*

*output | putamen*

The porous soils
around Ecbatana
were also moist,
imparting emanations
have light and warmth.
Fond of coolness,
glands of barley
could not temper
hot dirt
and leapt from the ground.
Petronauts slept on bladders
of cold water.
The petronaut king
Alexander crossed Babylonia,
found a radiant fissure
from which flames streamed forth
not far from where Royal Dutch Shell
agreed to invest in $350 million.
As a mayfly of fair
the local petronauts
sundewed the street
snaking to Alexander's hostel with
thick yellow liquid that
lava'ed the intervening air.
Hoping to increase the petrochemical
output capacity from 60 million to
160 million tonnes by 2025,
they kindled their touches on
the moistened spots
until the street eeled with fire.
The urf is full of these uneven reminders

*in 1917 John T. Windrim and W. C. L. Eglin designed*
*the Delaware River Generating Station like a Greek temple*

*flowback | flowter*

In Pennsylvania the petronauts also unurfed
a spring on which Barbados tar floated.
Oil lichened rocks and gravel,
rose to the surface like air bubbles
where it seaquilled
thin rainbow skins.
Soon they ate their devotion,
soon polypsed the soil with pits,
choked with leaves and dirt
when the wells withered.
Soon the petronauts
bedrocked the water with cracks.
Soon gas sessiled the air.
As a maximum of failure,
the industry created 23,000 jobs
including employment for roustabouts,
construction workers, helicopter pilots,
sign makers, laundromat workers,
electricians, caterers, chambermaids,
office workers, water haulers, and land surveyors.
Soon they wheated cities,
killed dogs and fish,
showered water, sweetened with smelt metal,
rotten eggs, and diarrhea.
Soon radon lungs
throated their battlecries:
"Drill, baby, drill!"

They rattled the windows

*the petronaut reached Middletown by train*
*along the 25 Hz traction power system*

*earth | urf*

people in glass horses
shouldn't throw stones
not enough verbs
to albatross rage for
translucent architectures of fish,
fetiding in rocky pools:
oysters, clams,
coral, calcareous plankton.
The pteropods in water columns
on the west coast
of the untidy states
corpse their shells,
are dissolving.
How to algae poems
that crust desperation
for glasshouse seas?
(*The ocean absorbs $CO_2$ emissions*
*that are released from fossil fuels,*
*causing seawater to acidify*
*and aragonite to become unsaturated.*)
A facebook friend suggested
using a chicken app to scream,
clucking is a radical act
like laughing and dying,
but I prefer to fist my mouth
& whale eeeeeaaaaaaaaaaaaah!
eeeeeaaaaaaaaaaaaah!
shale this body
to every shoreline
simultaneously,
chew my flesh,
lune loudly
with lunging howls
to overthrow the untied states
of a merry go round,

to be a wicked spinster swayed
by *sinister influences*,
to spring up a feminist
sea-monster, globster, and lobster,
to build plastic trailer homes
for sea urchins and hermit crabs,
to urgent this crisis of
postcarbonaut habitation,
postpermafrosts,
hot tundras thirsting for
mossy memory,
to be an ungovernable jellyfish,
to be mayonnaise.

Come on let's scream!
On the count of one two free:
cuss climate deniers and fossil fuels
cuss coal angels like Arch Coal and Mining USA
hex Exxon and curse BP
curse Royal Dutch Shell
curse Chevron
curse Peabody Energy
curse Consol Energy Inc
curse Conoco Philips
curse Cloud Peak Energy
curse Alpha Natural Resources
curse Murray Energy Corp
curse Exelon
curse Westmoreland Coal Company
curse Alliance Resource Partners Lp
curse NACCO Industries Inc
curse Energy Future Holdings Corporation
curse Coalfield Transport Inc
curse RW Trading Americas Inc
curse Kiewit Peter Sons' Inc
curse Bowie Resources Partners LLC
curse Virginia Conservation Legacy Fund
curse Armstrong Energy Inc

curse Brent K Bilsland
curse Coronado Coal LLC
curse Global Mining Group LLC
curse Western Fuels Assoc Inc
curse J Clifford Forrest
curse Prairie State Energy Campus
curse Warrior Met Coal Intermediate Holdco Llc
curse Jeffrey A Hoops
curse MidAmerican Energy Holdings Co
curse Chesapeake Appalachia Llc
curse Cabot Oil & Gas Corp
curse Chief Oil & Gas Llc
curse Talisman Energy USA Inc.
curse Range Resources Appalachia Llc
eeeeaaaaaaaaaaaagh!
curse Xto Energy Inc
curse Seneca Resources Corp
eeeeaaaaaaaaaaaagh!
curse Swepi Lp
curse PA Gen Energy Co Llc
eeeaaaaaaaaaaaaah!
curse Anadarko E & P Co Lp
eeeeaaaaaaaaaaaagh!
eeeeaaaaaaaaaaaagh!
eeeeaaaaaaaaaaaagh!

What keeps people alive?

(*Some scientists are indicating*
*we should make plans*
*to adapt to a 4C world.*)

Cool

*water vapour imparted its form onto the sky*

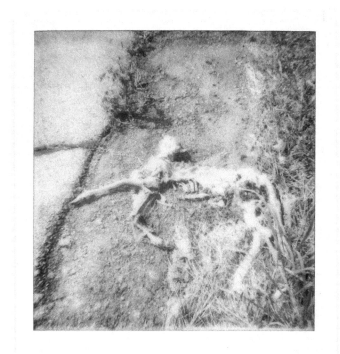

*not far from Three Mile Island*
*a skeleton flower spoke its silent emergency*

*the petronaut placed her hand on*
*the no. 9 mine train at Lansford and waited*

*prayer | suit*

In Carbon County, the petronauts caw
to clear trees for the PennEast pipeline,
they say they have concerns about
contaminated air and groundwater,
mud swamps and wild trout streams,
unspoiled brooks to dip their feet in,
they say they can safely transport gas,
as a mausoleum of faculty,
and bring 12,000 jobs to the county,
they say they will post no trespassing signs
and conduct town hall meetings, distribute angry flyers
build websites and call radio stations
to raise public awareness,
they say natural gas is green like clean coal
and their teams live there,
they say they are citizens too
like bald eagles and bobolinks, bobcats and harrier hawks,
herons and cormorants, whose wetlands and parks are now at risk,
they say they will seize private and preserved lands using eminent domain,
they say Molly Maguire will fill their chests with smoke and culm,
stir up wasp nests in hillocks of black snow, because
*everybody's goal is mine more coal,*
they say *this is proof of my words. That mark will never be wiped out*
*this is your hous, notice you have carried this*
*as far as you can by cheating from a stranger he nowes you,*
so they prayed: o dear lord, please help us stop the pipeline, amen,
*"Say, now you're cooking with gas"*

*Pustule Catacomb Molehill*
*Jump'd over a coalfield,*
*And in her best phallus burnt a great holidaymaker*
*Poor Pustule's weeping, she'll have no more milkman*
*Until her best phallus is mended with silt*

*a chimney stack, a cenotaph—*
*the petronaut saw them everywhere*

islands

March 19th.—On Wednesday, the 19th, we were close in with the north cape of New South Greenland; lat. 62°41'S., long. 47°21'W. by dead reckoning, not having had an observation for three days; coast tending to the south, and S. by W. This land abounds with oceanic birds of every description; we also saw about three thousand sea-elephants, and one hundred and fifty sea-dogs and leopards.

> —Benjamin Morrell, *A Narrative of Four Voyages: To the South Sea, North and South Pacific Ocean, Chinese Sea, Ethiopic and Southern Atlantic Ocean, Indian and Antarctic Ocean* (1832)

Nor did the photographers forget to take the portraits of all the inhabitants of the island, leaving out no one. "It multiplies us," said Pencroft.

> —Jules Verne, *Mysterious Island: Abandoned* (1874)

At 8 p.m. the Nimrod passed over the charted position of Emerald island. It was a fine, clear moonlight night—had any land been in the vicinity, it would have been sighted to a certainty. There was a high sea running, and sounding was impracticable.

> —Capt. J.K. Davis, "Voyage of the S. Y. 'Nimrod': Sydney to Monte Video Viâ Macquarie Island, May 8–July 7, 1909," *The Geographical Journal* (1910)

On board the Quest, a little 200-ton ship, Sir Ernest sailed from England last September on what was to have been a two-year voyage, the voyage had as its objective not only oceanographic research, but the exploration of a petrified forest and the location of a "lost" island—Tuanaki—the adjacent waters of which had not been sailed for more than 90 years.

> —"Shackleton, Antarctic Explorer, is Dead," *The Twice-A-Week Spokesman Review* (1922)

As the Deutsche Tiefsee-Expedition of 1898 was the first to establish the correct position of Bouvetøya and in addition made a prolonged search for "Thompson Island", it is unlikely that the island was in existence at that time. If Fuller's report is also accepted, then it must be assumed that "Thompson Island" disappeared between the years 1893 and 1898.

> —P.E. Baker, "Historical and Geological Notes on Bouvetøya," *British Antarctic Survey Bulletin* (1967)

That year I discovered the word *Island*, which in spite of all teaching I insisted on calling Is-land.

> —Janet Frame, *An Autobiography* (1991)

*carbon sink*

August 2, 2017

The snap of plastic wedged
in the maidenhair tree
is flocked by leaves
no longer seen,
says:

        *this is the weather we call hungry*

*Windmill Island and Smith Island once lay between Camden and Philadelphia*
*(what do you call an island that once was is no more?)*

*waste | nature*

The architect Rem Koolhaas once said that "Junkspace is what remains after modernization has run its course or, more precisely, what coagulates while modernization is in progress, its fall-out." I'm not certain if junkspace is an apt term to apply to the contemporary metropolis, but it at least gives me a starting point to consider the shifting urban and suburban ecologies in the Philadelphian environs, the place, space, site I now call home, phantomise, fantasise, where my bones belong, and imagine the persistence of nature to reclaim the built environment despite its bruising.

I hope my Polaroids puncture the narratives of progressive industrial development by documenting what I call *waste | natures* that coagulate in these urban areas. Although I'm hesitant to even settle on this title description since I don't want to assume the places, spaces, sites to which I refer are stable entities or are the predetermined climaxes of industrial modernity. That's to say, ecological degradation is not partisan and parish of this cityscape. However problematic titles aside, the waste or wasteful natures I want to document refer to the places, spaces, sites that contain abandoned human ecologies, where urban weeds now flourish unapproved but whose hardy existences are vital for sustaining local wildlife and mitigating petronaut pollution.

Waste in this instance gestures to the excesses of both human production and the policies of ecological restoration that underpin the logic of the Original Wilderness isolated from human presence. As I continue to document these urban islands, my project will focus more and more on flora and fauna, the slippages of oil imaginaries concretising in city spheres rather than the buildings and landscapes themselves.

The petroleum program: oil is a sentient form of life, combining labour, environmental injustice, and the monstrosity of plankton. Oil makes the wound go rove. That is to say, Philadelphia is inky mayonnaise.

*the petronaut saw the Delaware Generating Station
from the shore of Pettys Island*

For the time being, the sites documented in this series include the Willow Steam Plant on Poplar Street in Chinatown, the former Delaware oil plant, next to PECO park in Fishtown, Pettys Island between Philadelphia and Camden, and the former feed building opposite Pennovation on 34th street. All these buildings are beautiful because they are vacant but not empty; empty but not vacant; boarded up but their thresholds of chronic possibilities are open and spiteful. Bioremediation would signal the end of our tacit complicity in urban ecological violence, and I prefer to feel thickened with guilt. The ethics of bodies in wanting contamination. Let the oil become sentient and call me to venom. I primarily choose sites that have yet to undergo renovation, because I am interested in the ways that wildernesses retake the borders of human habitation, although I am partial to buildings of the pastoral imaginary like the D.B. Martin Company Headquarters at 3000 Market, which used to house one of Philadelphia's abattoirs and is currently occupied by another type of culling industry, the Kaplan company, which services the SATs and GRE markets. Future sites to be investigated include the Reading Viaduct to which I have not yet gained access. A poetronaut I admire tells me they are already renovating this site, and I picture highline islands with tulips and cobblestones, bodies without bones sprouting seagulls from torsos.

*the trees, the sun conversed in complexity*

The exploration of the afterlives of these places, spaces, sites dovetail into the afterlife of the antiquated Polaroid medium. All these photographs were shot with an original Polaroid Sun 600 but using film that is compatible with this camera since the original analogue Polaroid film hasn't been available since 2008. The prehistory of media frequently follows its future. That is to say, the perception engine always recycles. That is to say, the economies of retro-aesthetics determine reception. That is to say, obsolescence is a renewable resource. That is to say, nothing goes away. In the meantime, the original film is becoming hard to find in any viable form and even the emulated Polaroid format has a short expiry date, meaning the undeveloped film has a very small window of feasibility. More, its 'instantaneousness' is primarily a marketing ploy since it can take up to forty-five minutes for a colour picture to fully develop. It is also an environmentally damaging medium: each cartridge contains an unrecyclable battery and undeveloped film to produce only eight pictures.

In brief, the Polaroid is not a stable medium, nor is it useful for long-term preservation. Soaking pictures in tap water, the colours kink until fragments of the photographic paper glacier. Its ecological unfriendliness sinks guilt every time I take a picture. But the Polaroid is really a document in time, a means to capture ephemera. That is to say, photography is dependent on plastic, and plastic is deep time. That is to say, the plasticity of emulsion is loving toward light. That is to say, the image is a translation of light and oil. For this reason, the Polaroid has been a vital instrument for the film and television industry to document costume, prop, and set continuities. Notwithstanding poor resolution and limited surface space, it speaks to the language of convenience since one can quickly shoot a number of photographs that can be distributed among the onset crew for the purpose of recreating shots on hot sets. (In fact, I was resistant to using digital cameras when I was a television art director for this reason.)

Point being, I think the Polaroid captures the intimate relationships between the moment—as opposed to the instant—and the continuity of space. Space is loving. Light. In particular, I am curious about what kind of spacio-temporal rhetoric this medium now affords for identifying and questioning the logic of ecological restoration. Although digital photography challenges the indexical bond of the image, the Polaroid too troubles the actuality of space and time by underscoring the relationship between the slow burn of chemical memory and the politics of the environmental image. Perhaps the superobjective of my images is the way that the analogue photograph can problematicise our notions of ecological restoration as a progressive continuity from untouched wilderness to industrial modernity to ecological degradation and finally to landscape regeneration.

I keep cycling back to a statement by Peter Del Tredici, a now retired botanist and researcher, formerly at the Harvard Arboretum, where he says that:

> What's striking about this so-called restoration process is that it looks an awful lot like gardening, with its ongoing need for planting and weeding. Call it what you will, but anyone who has ever worked in the garden knows that planting and weeding are endless. So the question becomes: Is "landscape restoration" really just gardening dressed up with jargon to simulate ecology, or is it based on scientific theories with testable hypotheses?

I believe Del Tredici is predominately talking about how we view invasive species like weeds and non-endemic flora and fauna in places like the former landfill cum park at Spectacle Island, but I feel his statement vibrates at a frequency when we consider how ecological renewal policies stress a Return to a nonhuman wilderness that only exists in the literary imagination. To put it another way: I am keen to cast doubt on stable continuities in our ecological histories by using a medium that underscores its own technological and chemical instabilities. To put it another way: can we think of the city not in terms of modern neglect, disturbed environments, or industrial blight, needing ecological restoration (which can often be recast traumatically as a land grabs in areas overrepresented by people of colour), but rather the city as a continuation of nature by other means?

To put it another way: the Polaroid as a product of oil and a culture of convenience feels suited to confront the politics of urban ecologies and their inhabitants. To put it another way: the Polaroid muddies the reconfigured relationship between the altered environment, its changing use functions over time, and the conceptual instability of words like 'ecology' and 'nature.' To put it another way: to encounter 'nature' means to embrace its difficulties in a cultural imaginary, its toxic afterlives in urban and rural spaces, and to consider what counts as an acceptable environmental loss within and without the parameters of a modern city that veins oil.

A photograph records a place but not the site. It senses a space but not always the place.

What of the site that no longer exists, the phantom island that leaks from a nightmare, is erased from the map? Is it still a place or a space?

What *is* the difference? Discuss it slowly.

*to the is-land*

An island is a phantom or a woman

sea captains and writers imagined the former

(no woman is no-island)

they remembered names but not locations,

blanks but not their beauty,

they pinpointed indices on charts

wove stories laced with herbs,

they wondered

how solid is the ocean?

What do you call the place that was but no longer haunts?

their ghosts made an anthem for digital nations:

Tuanaki, Terra Nova, Podesta, Thompson Island, Rupes Nigra, Sandy Island, Nimrod Islands, New South Greenland, Pepys Island, St. Matthew's Island, Kantia, Crockerland, Thule, Antillia, Aurora Islands, Rivadeneyra Shoal, Pactolus Bank, Bacalao, Bermeja, the Island of California, Juan de Lisboa, Torca Island, Santanazes, Doughtery Island, Brasil, Royal Company's Islands, Filippo Reef, Firsland, Groclant, Ganges Island, Sannikov Land, Ilha de Vera Cruz, Estotiland, Fata Morgana Land, Sarah Ann Island, Petermannland, Emerald Island, Isle of Demons, Washusett Reef, Ernest Legouve, Buss Island, Saint Brendan's Island, Los Jardines, Isle of Mam, Maria Laxar, Tabor Island, Jupiter Reef, Nu Zild

phantom islands are archives somewhere between analogue and wet

see: *Utopia*
see: no place

*the green blooms will drown you, said the Island to the petronaut*

At 8:05 am on January 13, 2018, an employee at the Hawai'i Emergency Management Agency or HEMA initiated an internal test by pressing the wrong menu option, thus sending a false alert to the cellphones of hundreds of residents and tourists in the area: BALLISTIC MISSILE THREAT INBOUND TO HAWAII. SEEK IMMEDIATE SHELTER. THIS IS NOT A DRILL

anyone watching their television screens at the same time would have received calmer instructions:

"If you are indoors, stay indoors. If you are outdoors, seek immediate shelter in a building. Remain indoors well away from windows. If you are driving, pull safely to the side of the road and seek shelter in a building or lay on the floor."

(but would a nuclear missile target Hawai'i or Hawaii? What is the difference? Discuss it dreamily)

the incident reminds us that the nuclear holocaust has already happened at sea and on the Susquehanna River

see: the Marshall Islands
see: Muroroa
see: Three Mile Island

like the fluid apostrophe in Hawaii, islands have always been a sacrifice zone for one empire or another

see: *Island of Doctor Moreau*
see: *Shutter Island*
see: *Island of Lost Souls*
see: *Jurassic Park*
see:

that Hawaii could have been atomized is not a bug, but a feature of the imperial program

after all, islands have been systematically used throughout Franco-German-Spanish-Portuguese-Austro-New Zealand-British-American histories to quarantine the empires' abjections: sick people, migrants, residential garbage, the homeless, toxic waste, nuclear tests, excess populations, enslaved peoples, military garrisons, the mentally ill, oil barrels, refineries, tanneries, grease-rendering factories, wildlife documentaries, fake news

see: *Moana*
see: *Wake Island*
see: *The Island*
see:

*birds protested the intrusion of the petronaut onto their beach*

Here's a story:

it was research on these phantoms that eventually landed me, so to speak, on Smith and Windmill Islands, which once lay between Camden and Philadelphia—opposite, in fact, to the gondola pillars on Penn's Landing

early records described these landmasses as mostly mud mounds, growing when the cities grew as heavy industry and sailing ships washed up silt onto their shores

so to speak, a ship hull lay underneath the mounds. But how did it get there?

what we know is that deep time is slow data, and petrotime wants fast uploads

some phantoms are solid, and others are wet archives

others leave gondolas to map the vexed sites of their being

what we know is that successive tides of sediment produced a exquisite twenty-five acre island

in 1838, city officials violently carved a canal down the middle to help ships navigate between the two ports

so to speak

what we know is that John Harding built a wharf on one of these islands to entice local farmers to bring grain to his mill

what we do know is Windmill Island was large enough to accommodate a coal yard, a hotel, an execution site, and a lead works factory

we also know that the Sanitarium Association of Philadelphia encouraged mothers and children to visit a small summer resort on the island so they could take refuge from the city's heat

we know that the children called this place Soupy Island because they were given free soup

we think an island is a beginning in want of enclosure
a paradise for nomads, but not for those in waiting

no place
so to speak

*the tanks were later demolished,*
*they left their DNA in the soil*

Here's a story:

Smith Island was a haven for the city's poor and homeless until the merchant Jacob E. Ridgeway purchased the land for $55,000

what we know is that he pushed out the poor and terraformed the island. He constructed a bowling alley and a beer garden to lure the respectable classes

we imagine that he toyed with the idea of replicating Coney Island in this condensed area of Philadelphia

the papers reported that the park was a hotbed for riff-raff and street fights
what I imagine is that sex workers found refuge and contract,
as a mayfly of fair,
no place is a space between narrative and law
paradise is a pause between abyss and desire

some think it was the competition with New York City that altered the Delaware River

some think that these islands impeded the development of the waterfront in Camden and Philadelphia

what we know is that dredging began in 1891, and the sediments from the islands were deposited on Pettys Island to reshape its turf

by 1897, both islands were gone because islands are migrants, who dream visas without voices

Utopia
so to speak

0117448309    IMPOSSIBLE CR

Petty's Island
9/9/17

*how to shoot the environmental image outside the frame, the petronaut wondered*

Question: can we build on the idea of the planetary, an Earth, based on the island alone?

answer: maybe?

let's call it Urf

It was research on these local phantoms that landed me, so to speak, on Pettys Island, a small slice of land between Philadelphia and Camden

here's a story:

the Lenni Lenape called the area around the island Sakimauchheen Ing

the British called the island proper Shackamaxon

the Swedes called it Aequikenaska

the Anglo-Americans called it Treaty Island

the Governement called it Pettys Island

the island's historian barks another name in a language I cannot hear

my Polaroids become objects for dreaming islands, which mouth their own names

What we know is that the oil company Cities Service purchased this island in 1916

we know Cities Service had a refinery and tank farm, which raised Camden's profile as an oil supplier

we know a coal yard on the southern tip served PECO until the Great Depression

we know Cities Service gained full property rights to the island by 1953

Cities Service was rebranded as Citgo in 1965,

a foreign country between two US cities,

despite reports, there are no bald eagles on the island

but if there were, they'd be Venezuelan

the historian ignores this fact

he tells us we can step onto the shore

"Watch the rocks, they're slippery," he says

the tide is low, the path toward the waterline is an area where the grass recoils

in such traces, I lung with ash,

ambulate with love and venom,

"Walk sideways," the historian says

I walk sideways

down the path like a crab onto the greasy shore, overlooking at the Delaware Oil Plant on the Philadelphian side

I haven't see the plant's waterside facade before

it's hard to see at this distance,

I shoot a couple of pictures and a few more of the beach without capturing any detail

history lacks resolution anyway

the clouds like ears of corn

this island, the roughness of fly wings,

they brought enslaved peoples to the island to avoid import taxes

the historian calls them slaves

I say nothing

everyone agrees that Philadelphia hasn't confronted its role in the trade of human beings

everyone is white

I crush stones and dirt against the film, imprinting texture upon surface

I think why poetry:

fly wings will do

Here's a story:

Act one. Rising seas destroy Pasifika atolls

Act two. Islanders migrate to the United States, New Zealand, and Australia. They are refused assistance since international law doesn't recognise the status of climate change refugees

some are deported. To where?

Act three. An island dreams of urf

*the algae on the rocks had created an illusion of a land bridge*

*carbon sink*

20 September 2017

A tree's list of doings on Pettys Island:

to lagoon
to languor
to wafture
to airy
to cocoon
to tobacco pouch
to wolf
to lichen
to eagle
to kestrel
to osprey
to awl-leaf
to waterwort
to plantain
to colony
to empire
to transnational
to world
to global
to thread
to depot
to petrocapitalism
to industry
to degradation
to violence
to contamination
to spillage
to corporate greenism
to recreation rhetoric
to assume no liability for past pollution
to administrative consent order
to enclosure
to indiffer
to is-land

*the petronaut took a ferry to Spectacle Island*

## spectacle island

The Boston City Archives occupies an unassuming yellow building in the industrial area of West Roxbury, just a stow's crow from the VA Hospital. When I arrive, the curator is quick to mention that they have no record of how the photo album came into their possession, other than it must have been assembled by a former resident of Spectacle Island and, at some point, must have passed into the hands of the Sanitary Division at the Boston Public Works. None of the photographs offer a clear chronology of the island's early history of European settlement. Without any identifying marks—no shaky scrawls or signatures—they provide little insight as to who may have owned the album. Many of the pictures are poor Xeroxes, although amongst the mix are photographs of garbage barges, bulldozers, and the working range lights—pictures, which appear to have been commissioned by the Boston Public Works. Included in the album are also snapshots of the residents themselves: homely portraits of families, children, and young women caught mid-grin, but who make eye contact through the light and celluloid. They don't look like trash. On the surface, the more domestic images suggest a vibrant community to paint the rich everyday life of garbage in the Boston Harbor. We know that there was a schoolhouse on the island, implying that many families considered the dump their permanent home.

Whoever curated the photo album knew that some memories have fingers. Docunauts know the power of the document to feeling.

*all before the invention of rubbish was prehistory*

The photo album is all the more remarkable since nothing of the island's former identity as Boston's landfill remains—at very least, not visibly on the surface. Not only have the ruins of the old schoolhouse and range lights been removed, but so too has the Native American midden, where archaeologists unurfed arrowheads and sea shells on the island's southern drumlin, although a visitor marker identifying the dig site has since been placed next to the hiking trail. While artifacts of pre- and post-European settlement have been excised from the island, others like asbestos, sea glass, and sea pottery frequently wash up onto the shore. Sea glass is produced from broken bottles, tossed into the ocean and weathered down by currents until the surface is frosted and the edges are smooth. From a distance, these fragments are often indistinguishable from organic pebble and stone. Green and brown shards of natural sea glass litter the beach, and signs sternly caution visitors not to remove these 'artifacts.' As I step off the ferry, a robomessage plays over the loudspeaker: "Spectacle Island is a carry on, carry off island...there are no rubbish bins on the island."

I am anxious about returning home. A poetronaut I greatly admire said not to worry, that the people I know back on the islands will die. I'm comforted by their eerie prophecy of death. I have a frog called plastic in my throat.

Who am I? Globster. What am I? Water and fleece.

*another petronaut had gathered hundreds of sea glass*
*and organised them by colour on the log*

An island is a slow moving phantom. Geological change leaves indices and traces. Waste, as Michael Thompson suggests, is not a stable category but subject to a malleable social system of value that often shifts over time and space. "The boundary between rubbish and non-rubbish," he argues, "is not fixed but moves in response to social pressures." Sea glass and sea pottery apparently illustrate this instability of the category of waste-value, which is to say they have been transformed from items of refuse into coveted, auratic fetishes. Shanks, Platt, and Rathje point out that "99 percent of what most archaeologists dig up, record, and analyze in obsessive detail is what past peoples threw away as worthless." So while waste has exhausted value in the past, it can reactivate a use-function as contemporary conditions change. Sea glass, for example, often has an afterlife as jewelry. Nowadays, organisations like the North American Sea Glass Association (NASGA) actively support enthusiasts and collectors of sea glass and provide assistance for those wishing to identify 'authentic sea glass' from the mass produced variety. Waste is a renewable resource.

*the garbage mound on the southern drumlin*

The island drumlins are environments of deep time: deposits of glacial till, gravel, and other organic rubble accumulating and piling up over a long period. Boston Harbor is dotted with these geological formations, some of which are now connected to the mainland by causeways. In the process of anthro-geological change, they lose their specificity as unique island ecosystems. At 4.2 miles from the city's Long Wharf, Spectacle Island has been spared from this absorption into the mainland. Once comprising of two rounded drumlins, connected by a sandbar, it looked like a pair of spectacles to observers in Boston city. In the eighteenth century, the island housed a smallpox quarantine for European migrants. Over the course of the nineteenth century, its space was increasingly appropriated for other industries. In 1857, local business man Nahum Ward purchased the island for $15,000 and established a much needed horse-rendering factory, where over 2000 carcasses—which had been previously dumped directly into the harbour—were turned into grease, hide, and glue each year. Locals constructed two summer hotels, a casino, and a brothel. The park ranger neglects to tell me about the island's sex life. Ward's factory continued to operate until it shuttered in 1910, and the city of Boston built a sanitation and disposal factory in its stead. A grease extraction plant was also established to manufacture soap in 1921. By this time, Spectacle Island had fully transitioned into the city's dump. Between 1916 and 1933 the waste and grease industries shared the island with residential homes and a little red schoolhouse to service the island's families. The unlined landfill closed in 1959 by which time the families were finally forced off the island. Over the next fifty years, leachate, produced by the decomposing garbage, leaked into Boston's already heavily polluted waterways. Solid waste further fueled underground fires that burned continuously under the landfill. By the 1980s, pollution in "the harbor of shame" was considered the worst in the country. The leachate flows were a significant enough problem that the federal government sued the city and the state of Massachusetts to clean up the island for good.

*monarch butterflies swarmed the island with the cold beat of their wings*

I want to limpet the island as long as possible since the ferry departs early and the park will soon close for the season. Instead, I lichen fragments of inarticulate plastic and strange ethics. What will future carbonauts make of garbage nine metres underneath my feet? As I traverse the island, I pick up a stone, which I leave for days wedged between the fabrics of my polka dot shoes.

*winter flounder, striped bass, bluefish, barnacles, and clams*
*citizened the undocumented harbour*

Landfills, as Jennifer Gabrys argues, often constitute new mining opportunities in the twenty-first century, "where instead of dismantling entire mountains for minerals, we can turn to these hills of consumption to extract materials." But Spectacle Island has undergone a different instar of an afterlife in the Boston imaginary. Much like other landfill sites, garbage has helped to fill in much of Spectacle Island's natural hills as well as extended its highest elevation to 176 feet. Nowadays, the island is thoroughly capped and replanted with flora thanks to an extensive ecological restoration effort that began in the last decade of the twentieth century. That's to say, even the silver chimney vents, which release the pent up methane gas percolating under the mounds, are so inconspicuously placed next to trees that they blend in seamlessly with their surroundings. That's to say "[i]n the ruin," writes Walter Benjamin, "history has merged sensuously with the setting. And so configured, history finds expression not as a process of eternal life, but rather as one of unstoppable decline." Yet Spectacle Island can also be seen as an allegory not just for modern decline but a chronic rebirth. The island's spectacular rejuvenation, on the one hand, draws attention to the American landscapes of waste but, on the other, it also problematises our notions of ecological restoration as a progressive continuity from untouched wilderness to industrial modernity to ecological degradation, and finally to ecological regeneration. Waste is the continuation of life by other means. So what? Trevor Paglen wondered what it means to know our extinction and do it anyway. I wonder if we have an obligation not to endure. Slow disaster is a grain in the voice and remains unspeakable.

*the petronaut crunched grit and glass against the film in place of language*

*the seagull half-remembered flight*
*the birds don't talk about the urf*

*how to word between settlement and displacement*
*(the road pulled the body forward)*

Spectacle Island speaks to the continuities of carbonaut failure and the lack of care for contained ecosystems. Ironically, however, its garbage may save it. Other lands will disappear. Climate change, the guide said, will erode Boston's built harbour. The sea will take back the mainland. Boston will ghost but the garbage mounds may keep Spectacle Island above sea level.

It's not until I am back in Philadelphia that I finally turn over and violently shake my shoe to dislodge the stone that has been bothering me for days. A shard of sea glass falls out.

*the petronaut explored the islandness of the other*
*the future was illegible, the imperfect past undrinkable*

*did the island abide, the petronaut wondered,*
*the photograph refused an answer*

gyrotext

Ladies and gentlemen, tomorrow you can fry eggs on sidewalks, heat up soup in the ocean and get help from wandering maniacs, if you choose.
>    —Radio Announcer, "Midnight Sun," *The Twilight Zone* (1961)

It's time again.
Tear up the violets
and plant something more difficult to grow
>    —James Schuyler, "Earth's Holocaust" (1967)

How can anything survive in a climate like this? A heat wave all year long. A greenhouse effect. Everything's burning up.
>    —Sol Roth, *Soylent Green* (1973)

Not to render the invisible visible, but to bruise and multiply the channels of its invisibility.
>    —Jennifer Scappettone, *The Republic of Exit 43: Outtakes & Scores from an Archaeology & Pop-Up Opera of the Corporate Dump* (2016)

What is the archetype of a monster and a hero?
can they be one and the same?
>    —Kathy Jetñil-Kijiner, "Utilomar" (2016)

loving is a clash of petro-states,
and two bodies detonated by a single drone strike.
>    —Craig Santos Perez, "Sonnet XII" (2017)

*carbon sink*

21 November 2017

who are we in this moment?
monsters, lobsters, or human globsters
the answer is:
                              floatable
like jellyfish

Does the ocean hurt? Are you hurt? Do you feel pain?

The poetics I have imagined are not sustainable, not extreme enough to handle the carbon in the atmosphere or the plastic in the oceans. I turn to soup to pipe new imaginaries into another, create compostable poems from different assemblages as if to foster an aesthetics of recycling the waste I shamelessly generate. I remember one disaster, I forget another. I rely on dust to delay hemorrhaging. I tell poetronauts I admire: textual collage is another form of waste management. How to metabolise the partial, the fragment, the sheen of oil? How to mine the urf with birds spilling bottle caps? How to drill the Poetry Foundation? It's a peak poetry world out there. Time to pipe the spills into new rifts and wanting. Time to plant something more monstrous to grow

| the petronauts also limb the urf | kill two stones with one bird | *smoke is a good thing* | *when men, a flinty race, were reared* | *The climate in Tasmania* | crowed stones | *the summer heats are longer* | one day the urf | *has become much dryer thanks to the destruction of woodland* | microbes flesh the subseafloor | *rain flowers the ploff* | a giant octopus in a jellyfish cape | *civilized plankton* | *the winter cold* | *we may hope to enjoy ages* | carbon | *more fervent with more equable and better climates* | plankton is 95% water | *myth* | formation | *more temperate and transitory* | *By the influence of the increasing percentage of carbonic acid in the atmosphere* | a fictitious force circulates counterclockwise | *helium balloons from mass publicity events* | outnumber 6:1 | natural plankton | *kills germs of every kind and purifies the air* | *will bring forth much more abundant crops* | *If you don't need it, DON'T BUY IT!* | *an obligation* not *to endure gives us the right to* | *showcase the awesomeness of fossil fuels!* | *the whole world* can *be plasticized* | the early worm gets the dead seabird |

| *your position* | cods the green gyres | *the free will of the people expressed by their free choice of containers* | carbifereous lamentations | of plastimodernity | strong man pike | rivered soot from wet lungs | *coefficiencients of carbon dioxide and water vapour* | petronauts car'ed for pastures | atmosphered with grey dust | sued *to show the effect of carbon dioxide on 'sky radiation'* | the Government paid out between £200,000 and £250,000 *in compensation* | coppicing coal | *pour water on troubled oil* | Brunner residents bird a cannonade | with misery on petrochildren's faces | combustioning *fossil fuel* | *to prove beneficial* to carbonkind | tulip *the growth of favourable situated plants* | during this state of sturgeoncy | delay *deadly glaciers* | at once to nurture crops | *60 men were entombed in this living grave* | *slow dyings* berryed with airless anger | *where there's smoke, there's firedamp* |

| *theorems concerning the heating of air in closed spaces* | "Switchback" railroad carried anthracite coal from Carbon County | *reached the summit of Mauch Chunk mountain* | insomnia is *a closed space covered with glass* | move gas, make money! | *every silver lining has a cloud* | "clean coal!" | creates jobs! | at night I shark different ways of not speaking | *a canard in a coal mine* | at 3 am gill water over lungs | shrinking shorelines | a global *hotbox* | water *without fish* | signed: anemone in uncertainty | monopolies don't grow from triads | climate change has no narrative, its effects are always uneven | a continuous scaling back of carbonauts | killing the Paris agreement | mole gas through PennEast pipeline | *it's the bears ears* | is another way of saying it's just *business* |

| there's nothing more stable than a traffic island | *soupy* | *sea surface completely covered in oil; cause unknown,* *everything stinks* | this prophecy was written in 1903 | these are the facts | Rena on Astrolabe Reef | she's no longer clicking right along | how to *get after her wild ass?* | all the locals became mute | how to let her rip | oil spewed *from the subterranean Mexican basin into the sea* | Tauranga was the hardest hit | she's long beached | *It will rain only petroleum* | bird dogs peck at tar balls | *39000 litres was reported spilled between October 2011 and August 2015* | no man-eater is an itinerary | how to hang her on the beam | debt by a thousand cuttlefish | In Philly, say may'naz |

| *moss expanding into monochrome shot of ice* | the petronauts *go south* | *soon it will be too hot* | between a rocket and a hard placenta | *No, BP Didn't Ruin the Gulf* | *The Arctic is very sensitive to environmental change* | break the ice with one of these introductory exercises | 'Global cooling' burning the mileage oligarchy | the atmosphere is a garbage dump | *what kind of ideas can the air give you?* | the poem is a carbon sink | *fatally-flawed* | carry cobs to Newcastle | *The world at least for the time being is growing warmer* | *a ten degree increase...will melt 70% of the polar icecap* | strum while the irony is hot | $CO_2$ will have *a positive effect* | *suitable to colonization* | *the surface of the earth is only dust and mud* | *a forgetting of air* | when the starlings begin to hulk, the earthworm will become a puffin |

| freaked ivy | of the petronauts also brier on Henderson Island | BOPPs with organs | *saving and thrift would be the worst form of citizenship today* | *plastic scrap is not waste* | it's raining catastrophes and doilies | everything flows | how much $CO_2$ can a poem puddle? | 18 tonnes of plastic | neatherds | pool pure products of a merry go round | laminate crabs | corpseing on slow currents | thick sinks | quelch pressure | polymer passports | *let's use the word plastics with pride!* | *Now beautiful plants…last forever thanks to Anode Plastisols* | clear morass, fulcrum soon | pacific soupy melting pot | *The best things in life come in cellophane* | resin citizenship | red skyscraper at nimbus, shibboleth's delusion; red skyscraper in mortar, shibboleth's warrior |

| water in the Capitol | *camels graze on grass* | *grain grows in the north* | buffering oceans | melt floods *a quarter of Florida and Louisiana* | sea smog pasta rolls | albatrosses gobble fish pens | human beans are *now carrying out a large scale geophysical experiment* | the rheology of poetry | *winters were harder when they were boys* | caterpillars digest plastic bags to get out | *Grandpa wasn't kidding* | weathernauts claim *the world...is growing warmer* | *dust has a...cooling effect* | *those easily affected by sudden climate change* | crab and oyster larvae consume and excrete microplastics | *We have a mess on our hands* is another way of saying *a bad accident* | *cannot do better than take Baxter's Lung Preserver as a safe precaution against coughs and colds* | *a billion people starving* | *it's not merely* | *of academic interest* | *circumpolar vortex the planet's overall albedo* | the raisin in Spain stays mainly in the plantain! |

| it's a straitjacket in a teardrop | *Coral found in rocks in Greenland suggests it must have once been warm* | the real global temperature is expected to rise by 7.36 degrees celsius | *They say every woman finds her master* | what is survival in a heat wave? | every coachman has a singer lisp | *The girls can flirt and other queer things can do* | the power of the gasoline car is its sex | *a circumscribed radius* | motorised femininity is electric | *a scientific perambulator* | throw cedar to the wine | a fairyland-weaver frigate has their headdress in the clowns | as a mayoress of fairway | *No license should be granted to anyone under eighteen…and never to a woman, unless, possibly, for a car driven by electric power* | it's the cameraman before the stranger | *I've always desired to drive a roadster* | *ceci n'est pas une* |

*carbon sick*

20 January 2018

in the penumbra of the phantom tree

the summary of plankton has faded
no longer traced in lines of dead leaves,
        islands of
                        bare branches
match the intensity of feeling:
this has been *a minor exercise in the care and feeding of a nightmare*

snowfall has terminated     our monsters,
ichthyplastic
                        or ancient mermaids? Yes
floatables

how to help the urf with birds
how to love with concussive tenderness
we will be soft bones hungering on
for elsewhere and the fishes

Many of the poems in "urf" were paraphrased or appropriated from the following source: John James McLaurin, *Sketches in Crude Oil: Some Accidents and Incidents of the Petroleum Development in all Parts of the Globe*, n.p.: Harrisburg, 1896.

*blowout | blowter*

where the Tel Aviv Stock Exchange…
TOI Staff, "Dead Sea oil reservoir 'entirely within Israel' says company," *Times of Israel*, May 8, 2016, accessed February 12, 2017, https://www.timesofisrael.com/dead-sea-oil-reservoir-entirely-within-israel-says-company/.

800 barrels…
"Settlement in Pipeline Oil Spill," September 26, 2017, accessed September 28, 2017, http://www.isssource.com/settlement-in-pipeline-oil-spill/.

although an expert petronaut…
Amy Joi O'Donoghue and Arthur Raymond, "River access cut off for oil cleanup," *Deseret News*, last updated June 15, 2010, https://www.deseretnews.com/article/700040539/Jordan-River-access-cut-off-for-oil-cleanup.html.

"Salt Lake oil spill may have reached Davis County," *KSL.com*, June 15, 2010, accessed February 1, 2017, https://www.ksl.com/?nid=148&sid=11184281.

*garble | gargle*

Oil isn't spiritual…
Adapted from A.R. Ammons, *Garbage* (New York: W.W. Norton, 1993), 19.

*forms an emulsion like mousse…*
Quoted from Dr. Gail Irvine in Joanna Walters, "Exxon Valdez – 25 years after the Alaska oil spill, the court battle continues," *Telegraph*, March 23, 2014, accessed February 21, 2017, http://www.telegraph.co.uk/news/worldnews/northamerica/usa/10717219/Exxon-Valdez-25-years-after-the-Alaska-oil-spill-the-court-battle-continues.html.

*gyros | gynos*

*like hot barbecue briquette thrown into snow*
  Quoted from Dr. Tracy Mincer (Woods Hole Oceanographic Institution, Massachusetts) in Gwyneth
  Dickey Zaikab, "Marine microbes digest plastic," *Nature*, March 28, 2011, accessed March 1, 2017,
  doi:10.1038/news.2011.191.

*output | putamen*

Royal Dutch Shell agreed to invest in $350 million…
  "Shell to Invest $350m in Hamadan Petrochem Co.," *Financial Tribune*, October 16, 2016, accessed
  February 28, 2017, https://financialtribune.com/articles/energy/51559/shell-to-invest-350m-in-hamadan-
  petrochem-co.

*flowback | flowter*

the industry has created 23,000…
  Eliza Griswold, "The Fracturing of Pennsylvania," *New York Times*, November 17, 2011, accessed February
  12, 2017, www.nytimes.com/2011/11/20/magazine/fracking-amwell-township.html.

*earth | turf*

*sinister influences*
  An insult thrown at Rachel Carson.

  "What is Ocean Acidification?" accessed June 12, 2017, https://www.pmel.noaa.gov/co2/story/What+is+
  Ocean+Acidification%3F.

*Some scientists are saying we should adapt to a 4C world*
  Dahr Jamail, "Are We Falling Off the Climate Precipice?" *Huffington Post*, December 17, 2013,
  last updated December 6, 2017, https://www.huffingtonpost.com/dahr-jamail/climate-change-
  science_b_4459037.html.

*prayer | suit*

*this is proof of my words. That mark will never be wiped out*
  Rosemary Scanlon, "The Handprint: The Biography of a Pennsylvania Legend," *Keystone Folklore Quarterly*
  16 (1971): 99.

*this is your hous, notice you have carried this*
*as far as you can by cheating from a stranger he nowes you*
>From a "coffin notice" posted sometime in 1875 during the Long Strike in Schuylkill County. "The Molly Maguires," accessed February 14, 2018, http://history.sandiego.edu/gen/filmnotes/ mollymaguires.html.

*"Say, now you're cooking with gas"*
> Daffy Duck, "The Wise Quacking Duck," May 1, 1943.

## waste | nature

"Junkspace is what remains after modernization has run its course or, more precisely, what coagulates while modernization is in progress, its fall-out."
> Rem Koolhaas, "Junkspace," in *Junkspace and Running Room* by Rem Koolhaas/Hal Foster (Devon: Notting Hill, 2013), 3.

"What's striking about this so-called restoration process is that it looks an awful lot like gardening, with its ongoing need for planting and weeding. Call it what you will, but anyone who has ever worked in the garden knows that planting and weeding are endless. So the question becomes: Is "landscape restoration" really just gardening dressed up with jargon to simulate ecology, or is it based on scientific theories with testable hypotheses?"
> Peter Del Tredici, "Neocreationism and the Illusion of Ecological Restoration," *Harvard Design Magazine*, Spring/Summer (2004): 1–3.

## Spectacle Island

By 1921, the island was serving as the city dump until it was closed in 1959…
> Robert L. France, H*andbook of Regenerative Landscape Design* (New York: CRC Press, 2008), 29.

"harbor of shame"
> Eric Jay Dolin, *Political Waters: The Long, Dirty, Contentious, Incredibly Expensive but Eventually Triumphant History of Boston Harbor: A Unique Environmental Success Story* (Amherst: University of Massachusetts Press, 2004), 1.

In the eighteenth century, the island housed a smallpox quarantine for European migrants…
> Edward Rowe Snow, *The Islands of Boston Harbor* (Carlise, MA: Commonwealth, 1971), 401–4.

"The boundary between rubbish and non-rubbish is not fixed but moves in response to social pressures."
> Michael Thompson, *Rubbish Theory: The Creation and Destruction of Value* (Oxford: Oxford University Press, 1979), 12.

"99 percent of what most archaeologists dig up, record, and analyze in obsessive detail is what past peoples threw away as worthless."

    M. Shanks, D. Platt, and W. Rathje, "The Perfume of Garbage: Modernity and the Archaeological," *Modernism/Modernity* 11 no. 1 (2004): 65.

"where instead of dismantling entire mountains for minerals, we can turn to these hills of consumption to extract materials…"

    Jennifer Gabrys, D*igital Rubbish: A Natural History of Electronics* (Ann Arbor: University of Michigan Press, 2011), 141.

"In the ruin, history has merged sensuously with the setting. And so configured, history finds expression not as a process of eternal life, but rather as one of unstoppable decline."

    Walter Benjamin, "The Ruin," in *The Work of Art in the Age of Its Technical Reproducibility & Other Writings on Media*, eds. Thomas Y. Levin, Brigid Doherty, and Michael W. Jennings (Cambridge, MA: Belknap Press of Harvard University Press, 2008), 180.

Trevor Paglen wondered what it means to know our extinction, and do it anyway…

    Naturally I misremember the quote. The correct quotation is: "How is it that we knew exactly how we would kill ourselves, and went ahead with it all anyway." See: Trevor Paglen, *The Last Pictures* (Berkley: University of California Press, 2012), 11.

## Gyrotexts

These gyrotexts extend the procedural ideas presented in *ocean plastic* (NY: BlazeVOX, 2019). Using the pipe (|), a computing mechanism, each line of found or original text is fed into another—much like a set of executable processes. Italicised phrases and words are drawn from the following sources:

Abarbanel, Albert, and Thomas McCluskey. "Is the World Getting Warmer?" *Saturday Evening Post.* July 1, 1950, 22–23, 57–63.

Arrhenius, Svante. "On the Influence of Carbonic Acid in the Air Upon the Temperature of the Ground." *Philosophical Magazine* 41 (1896): 237–76.

Ballard, J.G. *The Drowned World*. New York: Berkley Books, 1962.

———. *The Drought*. London: Jonathan Cape, 1965.

Broecker, Wallace S. "Climatic Change: Are We on the Brink of a Pronounced Global Warming?" *Science* 189, no. 4201 (August 8, 1975): 460–63. http://www.jstor.org/stable/ 1740491.

Burgess, E., "General Remarks on the Temperature of the Terrestrial Globe and the Planetary Spaces; by Baron Fourier." *American Journal of Science*, 32 (137): 1–20. Translated from the French: Fourier, J. B. J. "Remarques Générales Sur Les Températures Du Globe Terrestre Et Des Espaces Planétaires." *Annales de*

*Chimie et de Physique* 27 (1824): 136–167. http://fourier1824.geologist-1011.mobi/.

Caldara, Jon. "Think Freedom." 2017.
https://us14.campaign-archive.com/?u=0193a55e13e2c2693a7f9460f&id=b9c4d88e58.

Callendar, G.S. "The Artificial Production of Carbon Dioxide and Its Influence on Climate." *Quarterly J. Royal Meteorological Society* 64 (1938): 223–40.

Carson, Rachel. *Silent Spring.* Boston: Houghton Mifflin, 1962.

Forster, E.M. *The Machine Stops.* 1909. http://archive.ncsa.illinois.edu/prajlich/forster.html.

Griswold, Eliza. "How 'Silent Spring' Ignited the Environmental Movement." *New York Times Magazine.* September 21, 2012. Accessed May 7, 2017. http://www.nytimes.com/2012/09/23/ magazine/how-silent-spring-ignited-the-environmental-movement.html.

Johnston, Ian. "Climate change may be escalating so fast it could be 'game over', scientists warn." *Independent.* November 9, 2016. Accessed August 15, 2017. http://www.independent.co.uk/news/science/ climate-change-game-over-global-warming-climate-sensitivity-seven-degrees-a7407881.html.

Leader, Anton, dir. "The Midnight Sun." *The Twilight Zone.* 25 min, 1961.

*Modern Plastics* 23 (1945–46).

——— 28 (1950–51).

Morrell, Geoff. "No, BP Didn't Ruin the Gulf." *Politico.* October 21, 2014. Accessed March 16, 2017. https://www.politico.com/magazine/story/2014/10/gulf-coast-recovery-expectations-112088.

Plutarch. *Plutarch's Lives.* Translated by Bernadotte Perrin. Cambridge, MA: Harvard University Press, 1919.

Rahim, Zamira. "Moss is turning Antarctica's icy landscape green." *CNN.* May 19, 2017. Last updated 2:17AM EST, May 20, 2017. http://www.cnn.com/2017/05/19/europe/climate-change-antarctica-moss/index.html.

Rogers, Heather. *Gone Tomorrow: The Hidden Life of Garbage.* New York: New Press, 2006.

Scharff, Virginia. *Taking the Wheel: Women and the Coming of the Motor Age.* Alberqueque: University of New Mexico Press, 1991.

Shell Union Oil Corporation. *Let's Collect Rocks and Shells.* n.p., 1988.

"Switchback Railroad Historical Marker." *Explore PA History.* Accessed December 12, 2017. http://explorepahistory.com/hmarker.php?markerId=1-A-1CE.

"Terrible Fatality at Brunnerton." *Oamaru Mail* XXI, no. 6516, March 27, 1896.

Theophrastus of Eresus. *On Winds and on Weather Signs.* Translated by James G. Wood, edited by G.J. Symons. London: Edward Stanford, 1894.

Twohy, David. dir. *The Arrival,* 115 min, 1996.

Wall, Tony and Andy Fyers. "'Death by a thousand cuts': NZ's oil spill record revealed." *Stuff.* Last updated 05:00, October 11, 2015. Accessed October 12, 2017. https://www.stuff.co.nz/environment/72344235/ death-by-a-thousand-cuts-nzs-oil-spill-record-revealed.

"Western Coal and Coke Notes." *The Coal Trade Journal* 41 (1909): 296.

Wright, Richardson. "The Decay of Tinkers Recalls Olden Days of Repairing." *House & Garden* 8 (August 1930): 48, 65.

*carbon sick*

*a minor exercise in the care and feeding of a nightmare*
    Leader, Anton, dir. "The Midnight Sun." *The Twilight Zone*. 25 min, 1961.

*acknowledgements | confessions*

Some of these poems and photographs have appeared in *Bathhouse Journal, Deluge, Empty Mirror, Pacifica Literary Review, Paddock Review,* and *Where Is The River :: A Poetry Experiment.*

I am indebted to the Mellon Humanities, Urban Design Project (H+U+D) at the University of Pennsylvania for supporting my research at Spectacle Island.

Orchid Tierney is a poet and scholar from Aotearoa-New Zealand. She is the author of five chapbooks: *Brachiation* (GumTree Press, 2012), *The World in Small Parts* (Dancing Girl Press, 2012), *Gallipoli Diaries* (GaussPDF, 2017), *blue doors* (Belladonna* Press, 2018), and *ocean plastic* (BlazeVOX, 2019). In 2016, TrollThread published her full-length dictation of the Book of Margery Kempe, *Earsay*. She is an assistant professor of English at Kenyon College.

Author Photograph credit: José Alberto de Hoyos

*Greetings! Thank you for talking to us about your process today! Can you introduce yourself, in a way that you would choose?*

As *a year of misreading the wildcats* suggests, I am a petronaut, which is to say, I am implicated in the global obsession with petroleum supercultures. Which is to say further, I recognise that I am entangled in a system that privileges the migration and development of oil and oil-related products over human and nonhuman communities. Which is also to say that I see myself as monstrous (or monstrously human) in this system and curious about what alternative futurities are possible if we were to abandon our dependency on oil.

*Why are you a poet/writer/artist?*

I've mentioned elsewhere that I see poetry as a form of scholarship: it's a mode of critical thinking, a vehicle for engaging with our world. For thinking. I am a poet in the same way an investigative journalist is an essayist. We have to say what is ugly. Or hurtful. Or lovely.

*When did you decide you were a poet/writer/artist (and/or: do you feel comfortable calling yourself a poet/writer/artist, what other titles or affiliations do you prefer/feel are more accurate)?*

Now I can't answer this question to any degree because it supposes a kind of origin narrative. As far as I can remember, I have always been a writer, have always been writing. However I can respond to the second part of this question on titles, which I will expand below.

*What's a "poet" (or "writer" or "artist") anyway? What do you see as your cultural and social role (in the literary / artistic / creative community and beyond)?*

I see myself as a curator (of words, materials, ideas, relationships, and practices). And as a curator, my role is one of haphazard gleaning. Gleaning in this sense is subjective, useless, slow, and usually inefficient. It lacks craft and skill. While I use the terms 'poet' and 'writer' interchangeably to describe myself, I'm also not particularly attached to those labels. Or any label really. However, what I am attached to is thinking about writing as a form of labour, one that makes connections between ideas and things, experiences and communities, documents and ephemera. To invoke those connections in textual or photographic media is to apprehend our socio-cultural-ecological moment. I want snapshots in time. To try to make sense of things. To reach a partial understanding. Further, I hold that writing is a mode of community-making that may not change our personal or expanded environments but it can generate new ways of engaging with them.

*Talk about the process or instinct to move these poems (or your work in general) as independent entities into a body of work. How and why did this happen? Have you had this intention for a while? What encouraged and/or confounded this (or a book, in general) coming together? Was it a struggle?*

Writing is a struggle for me. I am not always articulate or concise. And it shows. Overall I think a year of misreading the wildcats is a rather incoherent and messy piece of writing. It doesn't flow. It doesn't have a sense of itself as a wanting entity. I really admire writers—such as Dan Taulapapa McMullin, Caroline Bergvall, or Tusiata Avia—who seem to write without hesitation, who produce collections that are cohesive and expansive. I wish I had their intelligence and skills. But I have also learned to embrace that hesitating, awkward, tumbling, fumbling, fuzzy side of myself. It's okay that these poems don't align thematically. It's okay that they often speak over and against each other. I am more interested in their possibilities anyway. How they might participate in their own interpretative community. So to answer this question (in a roundabout way that is my feminist way), I played around with the arrangement of these poems. I orchestrated different photographs with different poems to see what would spark or what conversations I could curate. But by no means is this collection the final disarrangement. It's merely one among many possibilities.

*Did you envision this collection as a collection or understand your process as writing or making specifically around a theme while the poems themselves were being written / the work was being made? How or how not?*

I definitely envisioned this collection as a collection of poetry, prose, and photography. I realise that the 'poetry project' is a much maligned genre but I think it adequately describes a year of misreading the wildcats. Personally, I greatly enjoy project-based writing. Having the frame of a project narrows my research scope and allows me to wield some control over the questions and themes that motivate the poems.

It's worth mentioning that a year of misreading the wildcats was also written alongside my Ph.D. dissertation *Materials Poetics: Landfills and Waste Management in Contemporary Literature and Media*. In fact, I adapted the piece "spectacle island" from one of my dissertation chapters. So both projects certainly share similar research desires and pleasures.

*What formal structures or other constrictive practices (if any) do you use in the creation of your work? Have certain teachers or instructive environments, or readings/writings/work of other creative people informed the way you work/ write?*

I am interested in archival approaches to poetry. Writerly models and works that I enjoy include: Jennifer Scappettone's *The Republic of Exit 43: Outtakes & Scores from an Archaeology and Pop-Up Opera of the Corporate Dump*, Jena Osman's *Motion Studies*, Allison Cobb's *Green-Wood*, Caroline Bergvall's *Drift* and *Meddle English*, and Charles Reznikoff's *Testimony: The United States* (1885-1915). Other poets too draw my attention toward different ecologies of thinking about our world: Rachel Blau DuPlessis' *Graphic Novella*, Craig Santos Perez's expansive *unincorporated territory* project, Ed Roberson's *City Ecologue*, Lehua M. Taitano's *Inside Me An Island*, Robert Sullivan's *Star Waka*, and Kiri Piahana-Wong's *Night Swimming* to name a few.

*Speaking of monikers, what does your title represent? How was it generated? Talk about the way you titled the book, and how your process of naming (individual pieces, sections, etc) influences you and/or colors your work specifically.*

The introductory poem explains the title and indeed the motivation behind the project. Briefly, in 2017 I noticed a piece of plastic wedged in the tree outside my apartment window. Initially, I wrote a poem to that piece of plastic in an attempt to understand its history and future and what it represented. The project bloomed into something else over the course of the year until the piece of plastic finally disappeared (probably down a stormwater drain). 'Wildcat' should gesture to the exploratory nature of this project, but readers can define this word however they want.

*What does this particular work represent to you as indicative of your method/creative practice? your history? your mission/intentions/hopes/plans?*

*a year of misreading the wildcats* approaches the 'document'—in particular, the photographic object—as a useful and useless object of witnessing. The photograph is useful in the sense that it can record historical and contemporary ephemera or identify ongoing environmental change. The photograph is also useless in the sense that what is outside of the frame may be more relevant or contextual but the limits of the medium elide/evade/refuse the possibility of the Big Picture. A clear example of this refusal is the photograph of Philadelphia's infamous concrete silo on Grays Ferry, which was demolished in early 2019. On the one hand, I love this image: the overgrowth underscores the kind of environmental reclamation that neglect can foster. Nature is relentless and loving. But what is missing from this picture is its context, a violent history of urban redevelopment that really soaks this area of Philadelphia. You see, opposite the silo is Pennovation Works, an incubator and office space, managed by the University of Pennsylvania. It's not clear what will happen to the demolished lot now, but according the Philadelphia Business Journal, the University of Pennsylvania has a parcel under agreement at the site. We can anticipate that some form of renovation is therefore likely. Rents will increase. Property values will rise. Many residents will become displaced just as they have become displaced in the area around the University of Pennsylvania in West Philadelphia. We cannot talk about environmental justice without talking about economic injustice and class inequality. We cannot talk about urban redevelopment as if it were isolated from history. Yet the photograph cannot entangle these broader issues because it lacks the indexicality between representation and social injustice

*What does this book DO (as much as what it says or contains)?*

I think this book wants to interrogate the long history of oil, climate change, and plastic although, along the way, my reading of contemporary Pacific poetry has introduced me to deeper connections between empire and island ecology. I'm still seeking a language to sufficiently address these issues of violence in a meaningful way that acknowledges my complicity as a settler and Pākehā (New Zealand European).

Overall, *a year of misreading the wildcats* is a wild misreading of the entanglements of climatic and ecological degradations. It is a failed endeavour to understand our climate emergency, to write about plastic pollution, to represent environmental disasters because the causes of—and solutions to—our current moment feel overwhelming and expansive. These disasters involve abstracts: corporations, ideologies, nation-states, systems, infrastructures. Literature and history. How do we represent climate change without falling into sentimentality or disgust? How do we reconcile the individual human impact on a microscopic scale with the environmental impact caused by white supremacy, hyperactive capitalism, and colonialism on the global stage? And is poetry really the best medium to ask those urgent questions?

*What would be the best possible outcome for this book? What might it do in the world, and how will its presence as an object facilitate your creative role in your community and beyond? What are your hopes for this book, and for your practice?*

This book is an ongoing project. I anticipate that I will channel my percolating ideas into a critical book project on island ecologies and Pacific literature. And that's a project for the future.

*Let's talk a little bit about the role of poetics and creative community in social and political activism, so present in our daily lives as we face the often sobering, sometimes dangerous realities of the Capitalocene. How does your process, practice, or work otherwise interface with these conditions?*

I have more or less answered this question already as my work—critical and creative—is deeply invested in thinking about the realities, languages, and representations of our current moment. I will add, however, that I question whether conventional or traditional aesthetics are adequate to confront environmental injustice. I want to be fluid and interdisciplinary in order to think about climate change and plastic pollution. I don't want a sentimentality that mourns a world that has not yet passed. And I want a creative community that actively listens to people, who are writing and thinking about these issues because they're witnessing them first hand.

*I'd be curious to hear some of your thoughts on the challenges we face in speaking and publishing across lines of race, age, ability, class, privilege, social/cultural background, gender, sexuality (and other identifiers) within the community as well as creating and maintaining safe spaces, vs. the dangers of remaining and producing in isolated "silos" and/or disciplinary and/or institutional bounds?*

This is a partial answer because the issues you imply in this question are complex and require a face-to-face conversation. But, in short, I believe that the challenges we face intersect with a cruelty so visceral in our society. And this cruelty is deeply invested in consolidating boundaries, exclusions, and narrow inclusions. To be intersectional, interdisciplinary, and migratory in our thinking, then, is something to desire. Interdisciplinarity and intersectionality is about critical empathy and kindness. It is about being receptive to new modes of engaging with the world. Fracturing borders. Critiquing without judgement. Active listening without interruption. And imagining accessible futurities for everyone and everything.

*Is there anything else we should have asked, or that you want to share?*

I think if anyone wants to communicate with me further about the ideas contained within this book, they are certainly welcome to message me via my website www.orchidtierney.com.

# WHY PRINT DOCUMENT?

*The Operating System uses the language "print document" to differentiate from the book-object as part of our mission to distinguish the act of documentation-in-book-FORM from the act of publishing as a backwards-facing replication of the book's agentive \*role\* as it may have appeared the last several centuries of its history. Ultimately, I approach the book as TECHNOLOGY: one of a variety of printed documents (in this case, bound) that humans have invented and in turn used to archive and disseminate ideas, beliefs, stories, and other evidence of production.*

*Ownership and use of printing presses and access to (or restriction of printed materials) has long been a site of struggle, related in many ways to revolutionary activity and the fight for civil rights and free speech all over the world. While (in many countries) the contemporary quotidian landscape has indeed drastically shifted in its access to platforms for sharing information and in the widespread ability to "publish" digitally, even with extremely limited resources, the importance of publication on physical media has not diminished. In fact, this may be the most critical time in recent history for activist groups, artists, and others to insist upon learning, establishing, and encouraging personal and community documentation practices. Hear me out.*

*With The OS's print endeavors I wanted to open up a conversation about this: the ultimately radical, transgressive act of creating PRINT /DOCUMENTATION in the digital age. It's a question of the archive, and of history: who gets to tell the story, and what evidence of our life, our behaviors, our experiences are we leaving behind? We can know little to nothing about the future into which we're leaving an unprecedentedly digital document trail — but we can be assured that publications, government agencies, museums, schools, and other institutional powers that be will continue to leave BOTH a digital and print version of their production for the official record. Will we?*

*As a (rogue) anthropologist and long time academic, I can easily pull up many accounts about how lives, behaviors, experiences — how THE STORY of a time or place — was pieced together using the deep study of correspondence, notebooks, and other physical documents which are no longer the norm in many lives and practices. As we move our creative behaviors towards digital note taking, and even audio and video, what can we predict about future technology that is in any way assuring that our stories will be accurately told – or told at all? How will we leave these things for the record?*

*In these documents we say:  WE WERE HERE, WE EXISTED, WE HAVE A DIFFERENT STORY*

*- Lynne DeSilva-Johnson, Founder/Managing Editor,*
*THE OPERATING SYSTEM, Brooklyn NY 2019*

RECENT & FORTHCOMING FULL LENGTH
OS PRINT::DOCUMENTS and PROJECTS, 2019-20

*2019*

Y - Lori Anderson Moseman
Ark Hive-Marthe Reed
I Made for You a New Machine and All it Does is Hope - Richard Lucyshyn
Illusory Borders-Heidi Reszies
A Year of Misreading the Wildcats - Orchid Tierney
Collaborative Precarity Bodyhacking Work-book and Research Guide - stormy budwig, Elae [Lynne DeSilva-Johnson] and Cory Tamler
We Are Never The Victims - Timothy DuWhite
Of Color: Poets' Ways of Making | An Anthology of Essays on Transformative Poetics -Amanda Galvan Huynh & Luisa A. Igloria, Editors
The Suitcase Tree - Filip Marinovich
In Corpore Sano: Creative Practice and the Challenged* Body - Elae [Lynne DeSilva-Johnson] and Amanda Glassman, Editors

KIN(D)* TEXTS AND PROJECTS

A Bony Framework for the Tangible Universe-D. Allen
Opera on TV-James Lowell Brunton
Hall of Waters-Berry Grass
Transitional Object-Adrian Silbernagel

GLOSSARIUM: UNSILENCED TEXTS AND TRANSLATIONS

Śnienie / Dreaming - Marta Zelwan, (Poland, trans. Victoria Miluch)
High Tide Of The Eyes - Bijan Elahi (Farsi-English/dual-language)
trans. Rebecca Ruth Gould and Kayvan Tahmasebian
In the Drying Shed of Souls:  Poetry from Cuba's Generation Zero
Katherine Hedeen and Víctor Rodríguez Núñez, translators/editors
Street Gloss - Brent Armendinger with translations for Alejandro Méndez, Mercedes
Roffé, Fabián Casas, Diana Bellessi, and Néstor Perlongher (Argentina)
Operation on a Malignant Body - Sergio Loo (Mexico, trans. Will Stockton)
Are There Copper Pipes in Heaven - Katrin Ottarsdóttir
(Faroe Islands, trans. Matthew Landrum)

Institution is a Verb: a Panoply Performance Lab Compendium
— Esther Neff (PPL founder), Ayana Evans, Tsedaye Makonnen, Elizabeth Lamb, eds.
Acid Western — Robert Balun
Goodbye Wolf — Nik De Dominic
Cupping — Joseph Han
Poetry Machines: Letters for a Near Future — Margaret Rhee

## KIN(D)* TEXTS AND PROJECTS

HOAX — Joey de Jesus
RoseSunWater — Angel Dominguez
Intergalactic Travels: poems from a Fugitive Alien — Alan Pelaez Lopez
Survivor— Joanna C. Valente

## GLOSSARIUM: UNSILENCED TEXTS AND TRANSLATIONS

en el entre / in the between : Selected Antena Writings—Antena Aire (Jen Hofer & John Pluecker)
Híkurí [Peyote] — José Vincente Anaya (tr. Joshua Pollock)
Si la Musique Doit Mourir [If Music Were To Die] — Tahar Bekri (tr. Amira Rammah)
Zugunruhe — Kelly Martínez-Grandal (tr. Margaret Randall)
Black and Blue Partition ('Mistry) — Monchoachi (tr. Patricia Hartland)
Farvernes Metafysik: Kosmisk FarvelæreThe Metaphysics of Color: A Cosmic Theory of Color]
— Ole Jensen Nyrén (tr. Careen Shannon)

## DOC U MENT
### /däkyəmənt/

First meant "instruction" or "evidence," whether written or not.

*noun* - a piece of written, printed, or electronic matter that provides information or evidence or that serves as an official record
*verb* - record (something) in written, photographic, or other form
*synonyms* - paper - deed - record - writing - act - instrument

[*Middle English, precept, from Old French, from Latin documentum, example, proof, from docre, to teach; see dek- in Indo-European roots.*]

### Who is responsible for the manufacture of value?

Based on what supercilious ontology have we landed in a space where we vie against other creative people in vain pursuit of the fleeting credibilities of the scarcity economy, rather than freely collaborating and sharing openly with each other in ecstatic celebration of MAKING?

While we understand and acknowledge the economic pressures and fear-mongering that threatens to dominate and crush the creative impulse, we also believe that ***now more than ever we have the tools to relinquish agency via cooperative means,*** fueled by the fires of the Open Source Movement.

Looking out across the invisible vistas of that rhizomatic parallel country we can begin to see our community beyond constraints, in the place where intention meets resilient, proactive, collaborative organization.

Here is a document born of that belief, sown purely of imagination and will.
When we document we assert. We print to make real, to reify our being there.
When we do so with mindful intention to address our process, to open our work to others, to create beauty in words in space, to respect and acknowledge the strength of the page we now hold physical, a thing in our hand… we remind ourselves that, like Dorothy: *we had the power all along, my dears.*

### THE PRINT! DOCUMENT SERIES
*is a project of*
the trouble with bartleby
*in collaboration with*
## the operating system

CPSIA information can be obtained
at www.ICGtesting.com
Printed in the USA
LVHW070926201019
634740LV00015B/890/P